ACCIDENTAL
SCIENCE DISCOVERIES

MICROWAVES

Kenny Abdo

Fly!
An Imprint of Abdo Zoom
abdobooks.com

abdobooks.com

Published by Abdo Zoom, a division of ABDO, P.O. Box 398166, Minneapolis, Minnesota 55439. Copyright © 2024 by Abdo Consulting Group, Inc. International copyrights reserved in all countries. No part of this book may be reproduced in any form without written permission from the publisher. Fly!™ is a trademark and logo of Abdo Zoom.

Printed in the United States of America, North Mankato, Minnesota.
102023
012024

THIS BOOK CONTAINS RECYCLED MATERIALS

Photo Credits: Alamy, Getty Images, Shutterstock
Production Contributors: Kenny Abdo, Jennie Forsberg, Grace Hansen
Design Contributors: Candice Keimig, Neil Klinepier, Colleen McLaren

Library of Congress Control Number: 2023938022

Publisher's Cataloging-in-Publication Data

Names: Abdo, Kenny, author.
Title: Microwaves / by Kenny Abdo
Description: Minneapolis, Minnesota : Abdo Zoom, 2024 | Series: Accidental science discoveries | Includes online resources and index.
Identifiers: ISBN 9781098284114 (lib. bdg.) | ISBN 9781098284831 (eBook) | ISBN 9781098285197 (Read-to-Me eBook)
Subjects: LCSH: Microwave ovens--Juvenile literature. | Microwave heating--Juvenile literature. | Serendipity in science--Juvenile literature. | Inventions--Juvenile literature. | Discoveries in science--Juvenile literature.
Classification: DDC 500--dc23

TABLE OF CONTENTS

Microwaves 4

The Accident 6

The Discovery 10

The Footprint 20

Glossary 22

Online Resources 23

Index 24

MICROWAVES

No one knew that tinkering with **World War II radar** systems would lead to one of the most common household **appliances**, the microwave!

THE ACCIDENT

In 1945, **engineer** Percy Spencer noticed something interesting while working on a **radar** set. A chocolate bar in his pocket had melted!

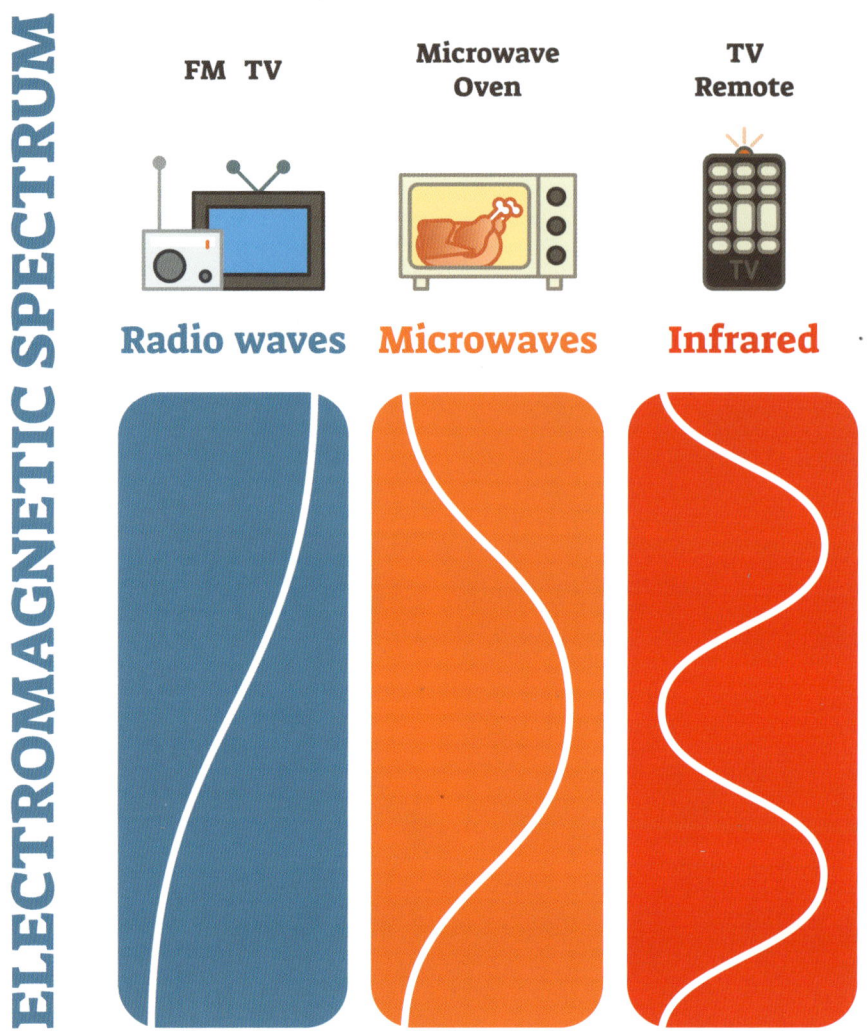

Spencer then began experimenting with different foods. He placed popcorn kernels next to the set. Soon, they began to pop!

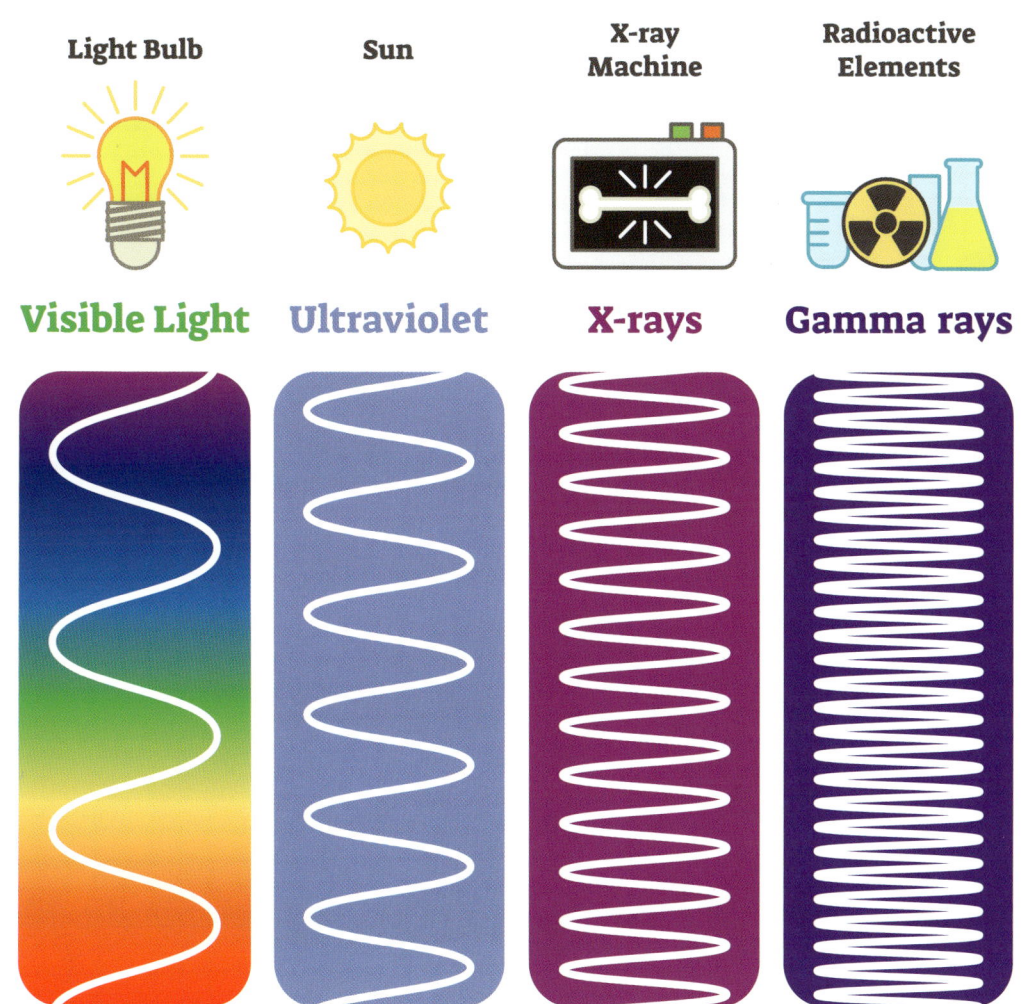

After a few more tests, Spencer found that food could also be cooked using **radio waves**. His idea was beginning to heat up!

THE DISCOVERY

In 1947, Spencer filed a **patent** for the first microwave. He called it the Radarange. It was very expensive and weighed more than 700 lbs (317.5 kg). The Radarange was mainly used in restaurant kitchens.

Golf Champion Julius Boros and his wife Armen have just made the greatest cooking discovery sinc

The incredible new *Amana Radarange* Microwave Oven. It bakes a p 4 minutes, sizzles a hamburger in 60 seconds, cuts most cooking tim

Golf Champion Julius Boros is as much at home in the kitchen as he is on the course. That's why there's an Amana Radarange Oven in the Boros home. It's the perfect answer for the large sports-oriented family. Here's what the Radarange Oven will do for you (and what it does for the Boros family):

1. Work miracles of speed—do a 5-pound rolled roast in 37 minutes. Roast a turkey in ¼ the usual time.
2. Reheat leftovers like fresh food.
3. Cook cool—no hot kitchen. Only the food being cooked gets warm.
4. Defrost frozen food in a jiffy—1½ min tes for 10 ounces of frozen fruit, 2 minutes per pound for roasts.
5. Make diet cooking easier—it's a cinch to "avoid fried foods," with

Others, like bacon, can be cooked on pa to absorb excessive fat.

6. Prepare fresh hot meals for "late no more dried-out dinners or food for people who work lat
7. Help you entertain—c appetizers, hamburgers e right on the table.
8. Help you avoid mess —you cook on paper, plastic. The oven wipes a damp cloth.

Plug it in anywhere, th Radarange Oven operate dard 115-volt outlets, use electricity than a frypan. on the kitchen counter (jus 22¾" wide, 17¼" deep ov your Amana Dealer, or

Amana Radarange MICROWAVE OVEN

Backed by a century-old tradition of fine craftsmanship.
Amana Refrigeration, Inc. Amana, Iowa, Subsidiary of Raytheon Company

By the 1950s, the first home microwave was introduced. It was not very popular due to its size and price. When the cost to make them lowered in the 1960s, more people were able to afford them.

By the 1970s, microwaves had become a staple in American homes. Cookbooks dedicated to microwave cooking also took off. Their popularity continued to grow throughout the world.

MICROWAVE COOKING

St Michael

JENNY WEBB

Today, microwaves are a common **appliance** found in most households. They are used to heat up leftovers, cook frozen dinners, and even prepare fresh meals.

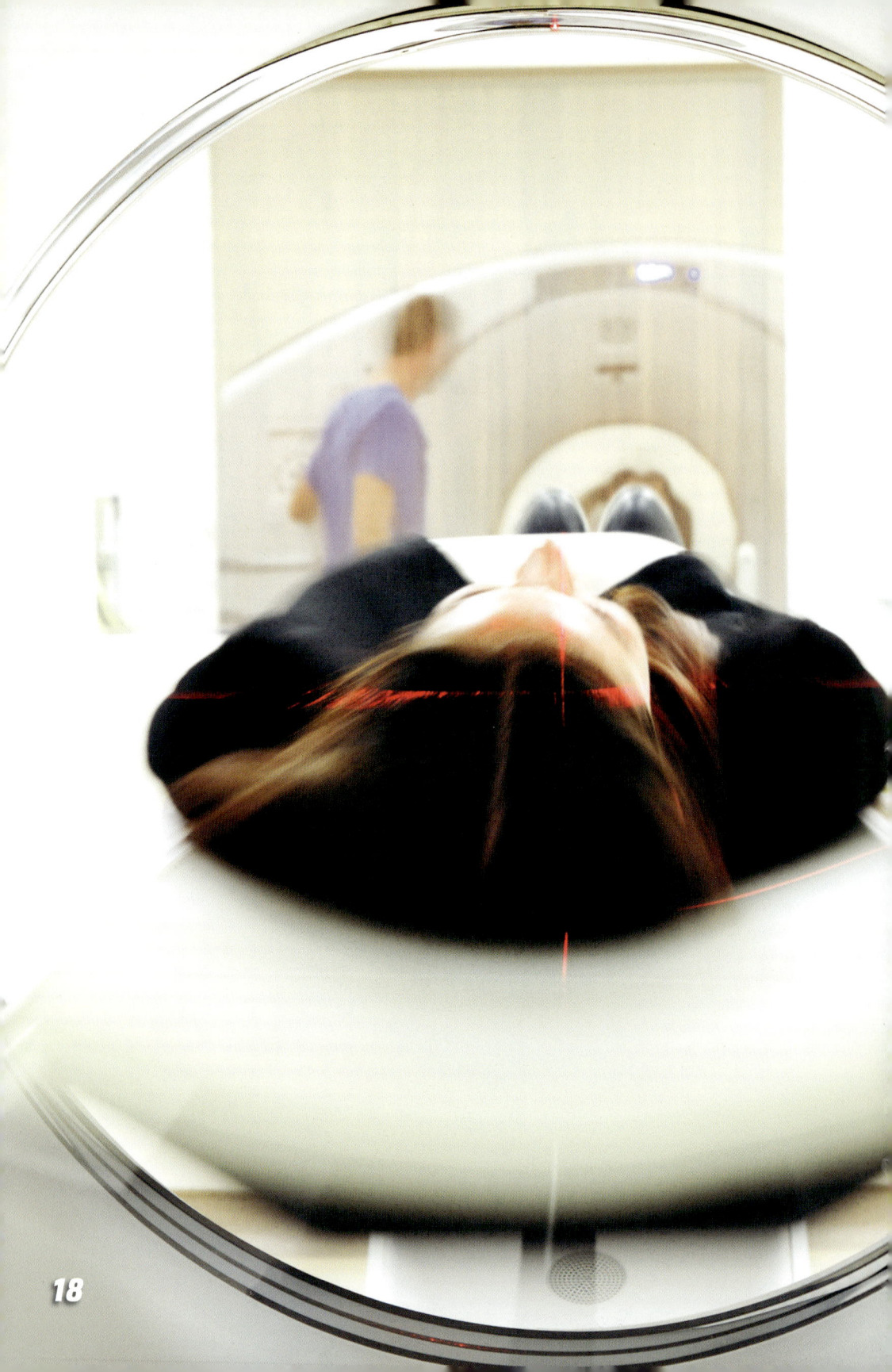

Microwaves are also used outside of cooking. They can send information over long distances. They are used to run **MRI** machines. They are also used to study the universe.

THE FOOTPRINT

Microwaves have changed the way we prepare food. They continue to cook up new scientific and technological discoveries!

GLOSSARY

appliance – a household device operated by gas or electric current such as stoves, refrigerators, and dishwashers.

engineer – a person who creates or builds structures and devices using science and math.

MRI – short for magnetic resonance imaging. A test that uses magnets and radio waves to create detailed pictures from the inside of the body.

patent – the exclusive right granted to a person to make or sell an invention. This right lasts for a certain period of time.

radar – an instrument that uses the reflection of radio waves to detect and track objects.

radio waves – a type of electromagnetic radiation with the longest wavelengths in the electromagnetic spectrum.

World War II – (1939-1945) a war fought in Europe, Asia, and Africa. Great Britain, France, the United States, the Soviet Union, and their allies were on one side. Germany, Italy, Japan, and their allies were on the other side.

ONLINE RESOURCES

To learn more about microwaves, please visit **abdobooklinks.com** or scan this QR code. These links are routinely monitored and updated to provide the most current information available.

INDEX

astronomy 19

cookbooks 14

MRI machines 19

Radarange 10, 13

radio waves 9

Spencer, Percy 7, 8, 10

uses 10, 14, 16, 19, 21

World War II 5, 7